Reduce

by Ann-Marie Kishel

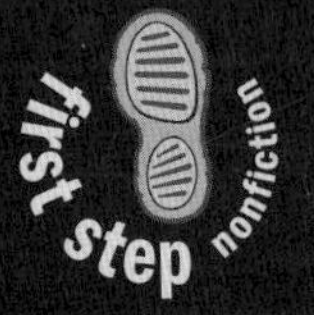

Lerner Publications Company · Minneapolis

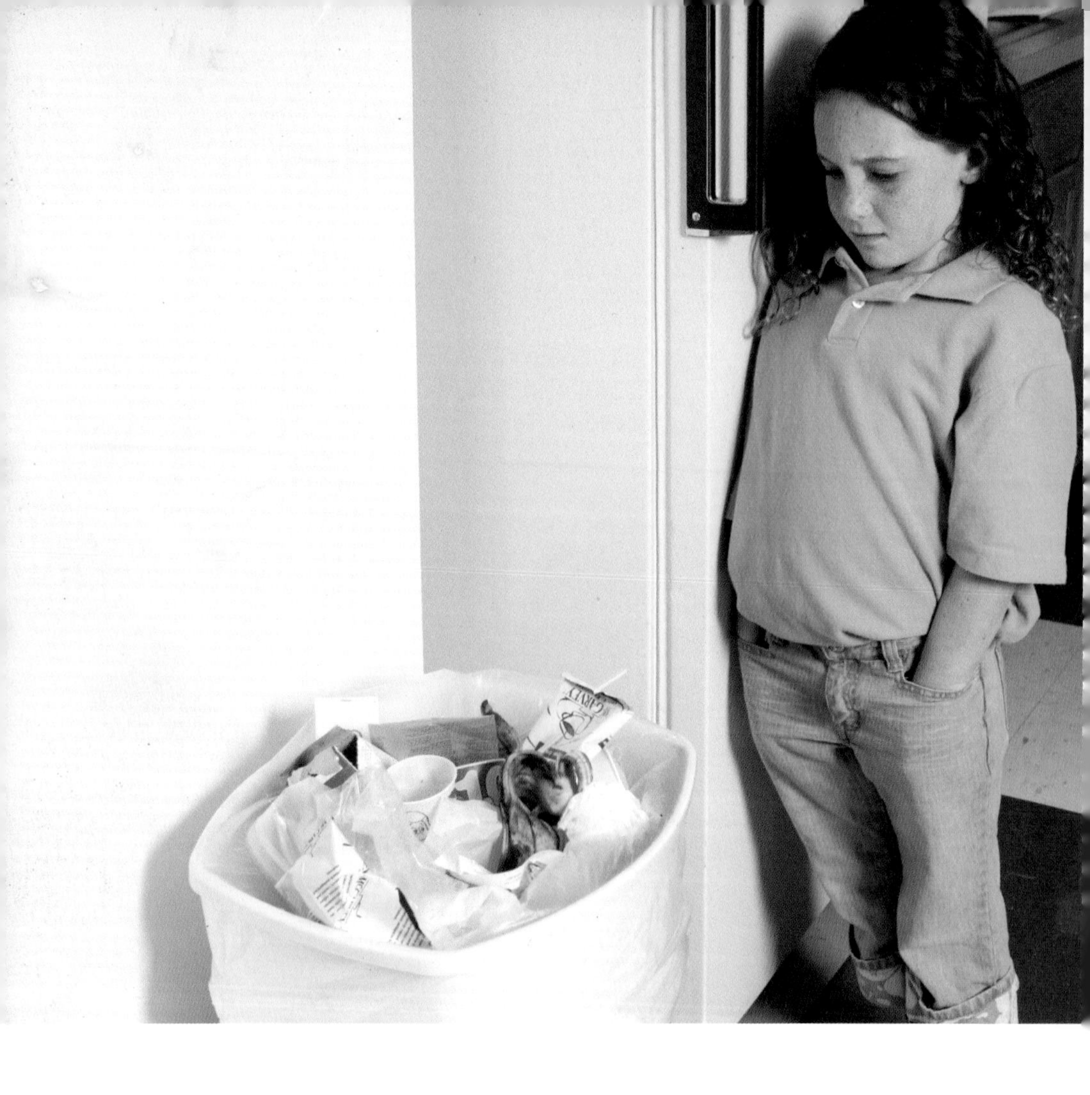

How can you reduce?

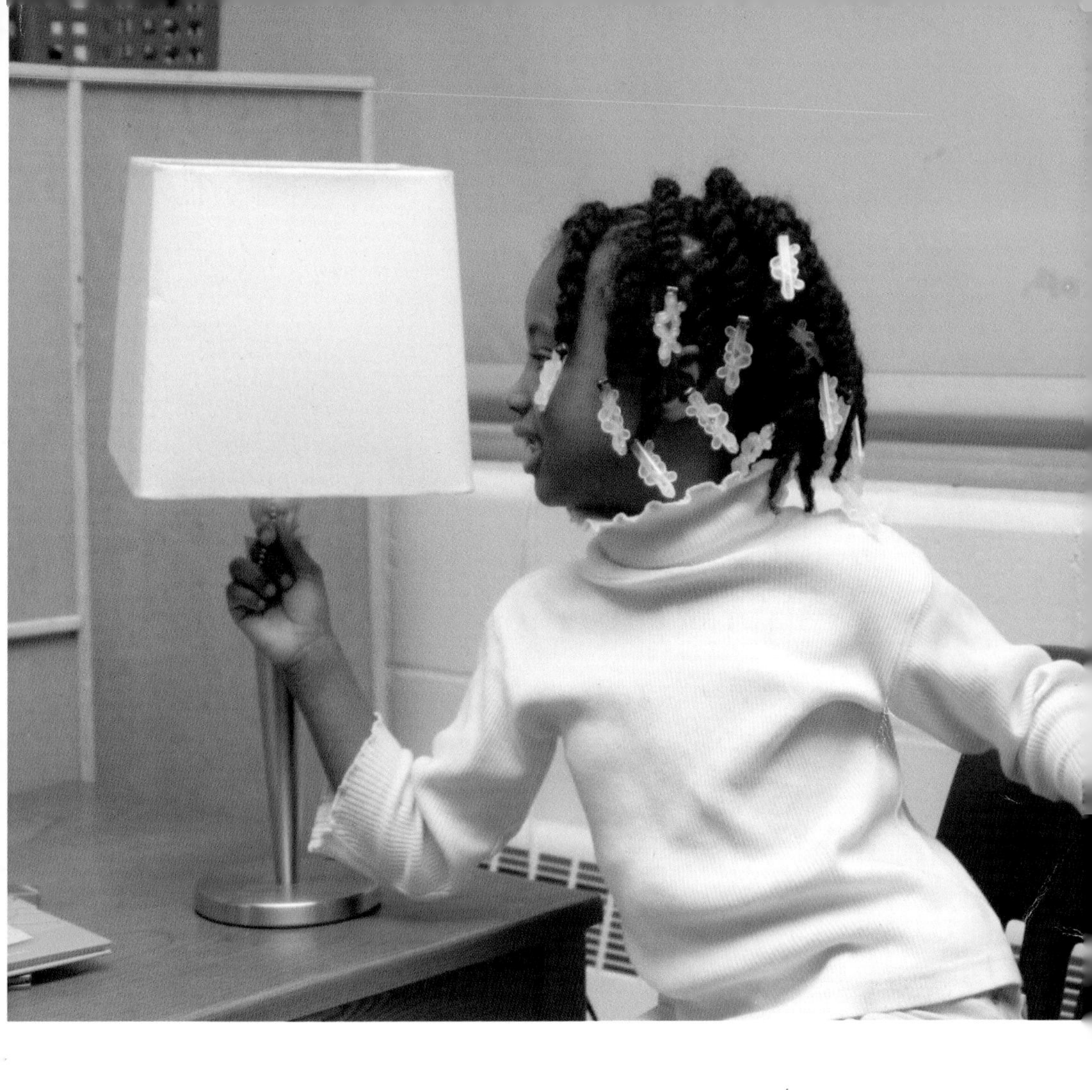

Turn the lights off.

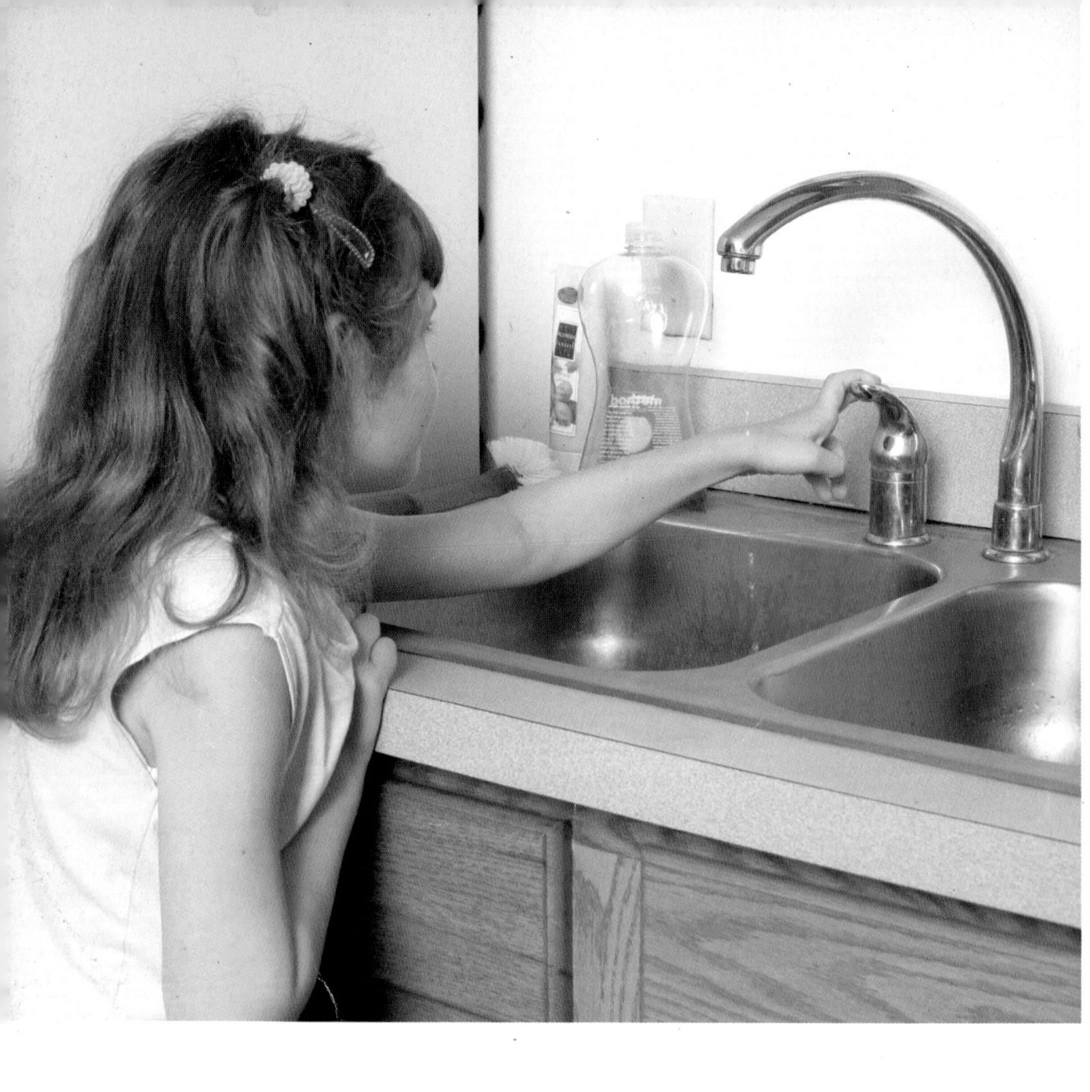

Turn the water off.

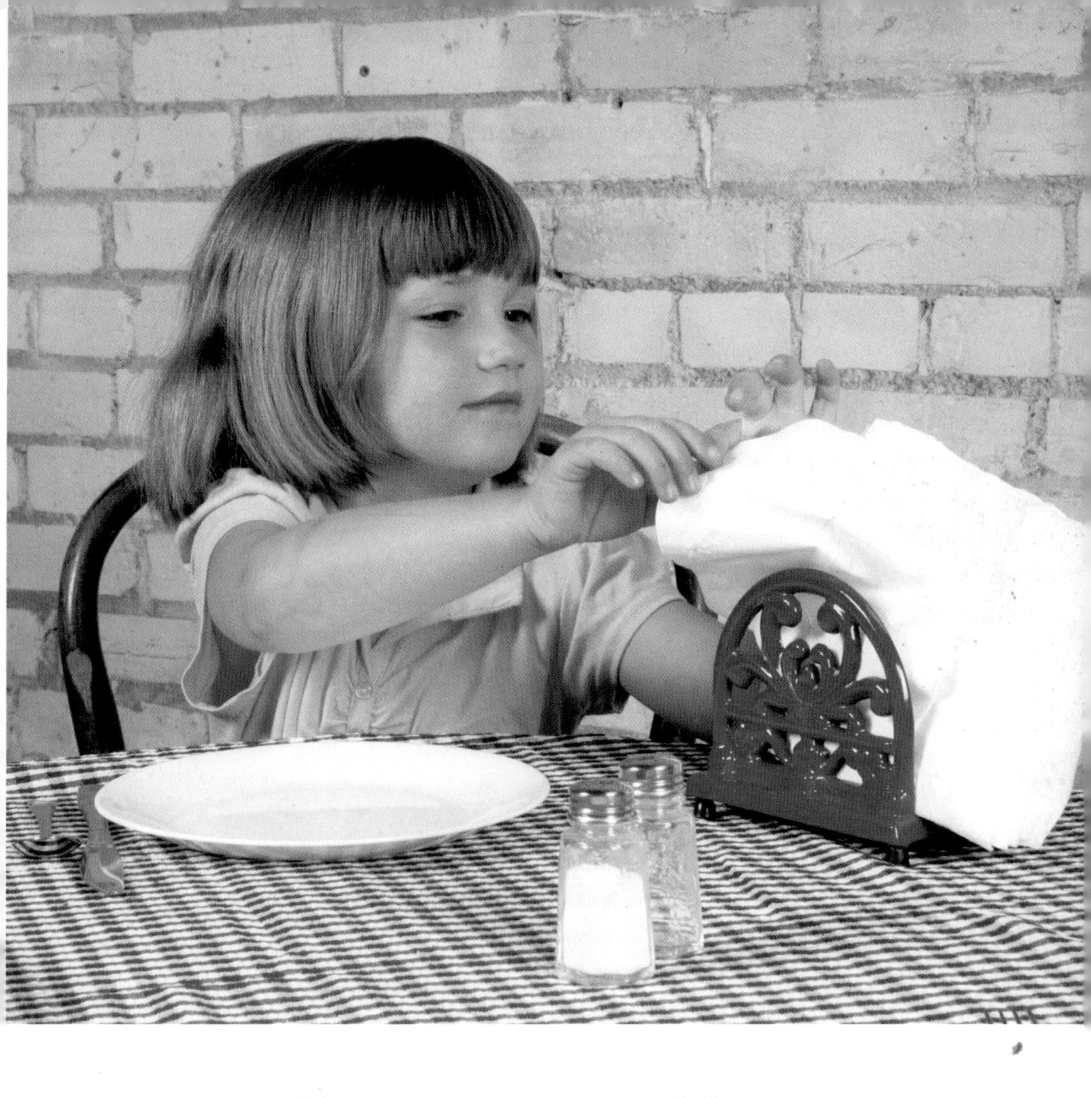

Use one napkin.

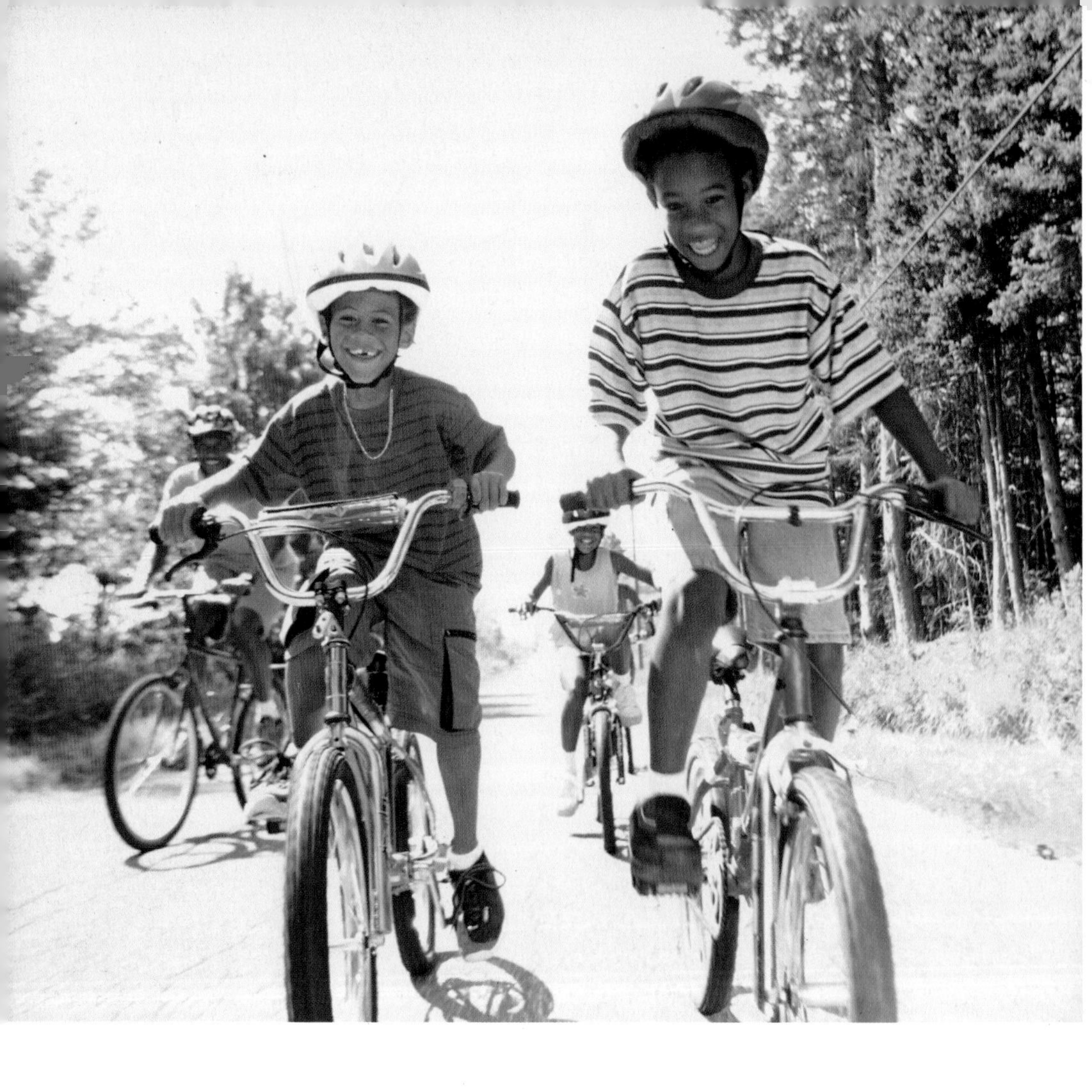

Ride a bike.

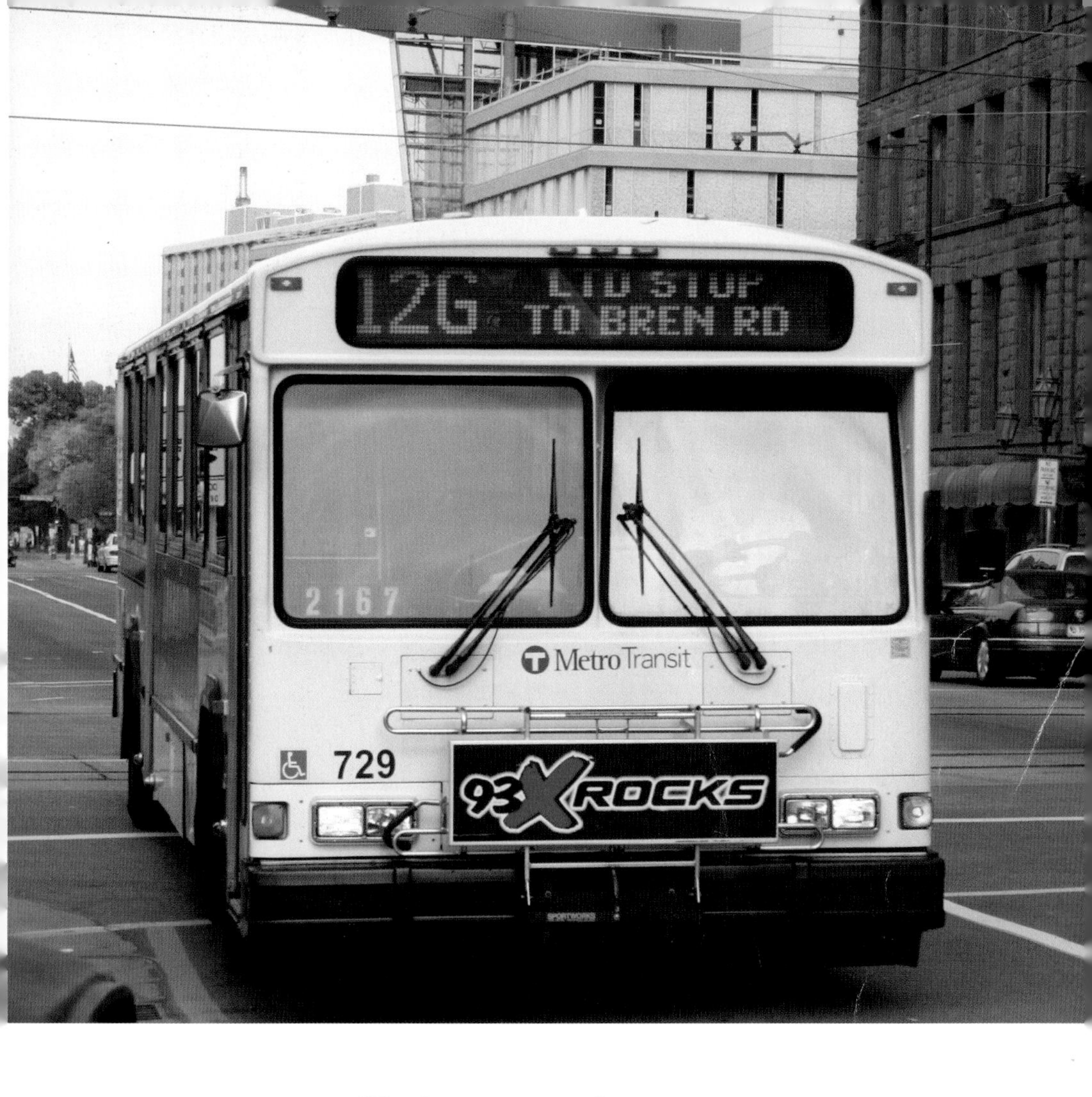

Take a bus.

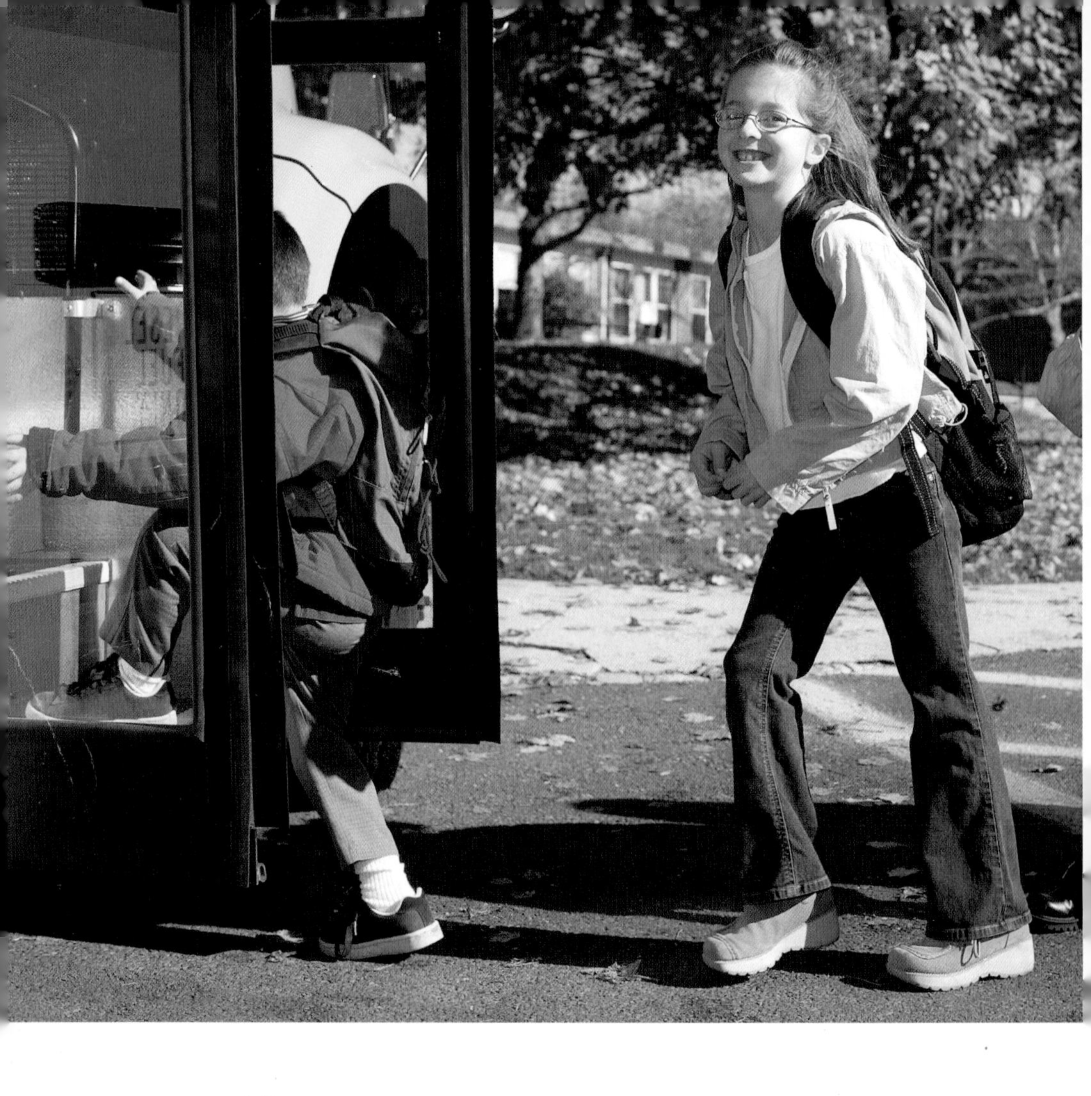

There are many ways you can reduce.